LET'S FIND OUT! EA

WHAT IS A LANDFORM?

LOUISE SPILSBURY

Britannica
Educational Publishing

IN ASSOCIATION WITH

ROSEN
EDUCATIONAL SERVICES

Published in 2014 by Britannica Educational Publishing (a trademark of Encyclopædia Britannica, Inc.) in association with The Rosen Publishing Group, Inc.
29 East 21st Street, New York, NY 10010

Copyright © 2014 by Encyclopædia Britannica, Inc. Britannica, Encyclopædia Britannica, and the Thistle logo are registered trademarks of Encyclopædia Britannica, Inc. All rights reserved.

Rosen Publishing materials copyright © 2014 The Rosen Publishing Group, Inc. All rights reserved.

Distributed exclusively by Rosen Publishing.
To see additional Britannica Educational Publishing titles, go to rosenpublishing.com.

First Edition

Britannica Educational Publishing
J.E. Luebering: Director, Core Reference Group
Mary Rose McCudden: Editor, Britannica Student Encyclopedia

Rosen Publishing
Hope Lourie Killcoyne: Executive Editor
Nelson Sá: Art Director

Library of Congress Cataloging-in-Publication Data

Spilsbury, Louise.
What is a landform?/Louise Spilsbury.
  p. cm. — (Let's find out: earth science)
Includes index.
ISBN 978-1-62275-257-7 (eBook)
1. Landforms—Juvenile literature. I. Spilsbury, Louise. II. Title.
GB406.S65 2014
551.41—dc23

Photo credits
Cover: Shutterstock: My Good Images. Inside: Dreamstime: Agap13 4, Americanspirit 7, Dejan750 22, Georgeburba 5, Helderpc 28, Jiajianzheng 8, Minute 14, Shargaljut 11, Silksatsunrise 6–7, Tihov 13, Timhesterphotography 10, Wisconsinart 9; Shutterstock: Denis Burdin 20, Caminoel 26–27, Jo Crebbin 15, Zack Frank 12–13, Jason Ho 27, Joyfull 16, Peter Kunasz 19, Mike Lane 24, Martin M303 18, My Good Images 1, Maxim Petrichuk 21, Tracing Tea 17, Amy Tseng 23, Krzysztof Wiktor 29, Ian Woolcock 25.

Contents

Landforms of the World	4
Plains	6
Plateaus	8
Mountains	10
Hills	12
Glaciers	14
Valleys	16
Canyons	18
Sand Dunes	20
Caves	22
Coastlines	24
Islands	26
Changing Landforms	28
Glossary	30
For More Information	31
Index	32

Landforms of the World

Nearly three-quarters of Earth's surface is covered with water (mostly oceans). The rest of Earth's surface is covered in land that is incredibly varied. It ranges from huge, mighty mountains to dark caves deep underground, from hills of sand in the desert to islands. These and other features, shapes, and types of land are called landforms.

Earth has some weird and wonderful landforms, such as this rocky landscape in Bryce Canyon National Park, Utah.

>> The force of a rushing waterfall can gradually break down and wear away rock.

Landforms vary around the world. Australia has large areas of sandy desert, and Nepal has many mountains.

Most landforms form slowly, for example, when rock is gradually worn down by erosion. Some landforms form more quickly, for example, when hot lava from underground bursts out onto the surface and hardens.

Erosion is when soil or rock is carried away by water, ice, or wind.

Plains

Plains are wide, flat, nearly treeless areas of land surrounded by areas of higher land. These common landforms cover around one-third of all land on Earth. They are found on all continents except Antarctica.

Many plains are covered by grassland. These are areas where mainly grass grows and animals such as zebra, deer, and bison live.

Compare and Contrast
Many plains animals have hooves. Compare and contrast hooves and other shapes of feet. How are hooves useful for life on the plains?

▶▶ East African plains are covered in coarse grasses.

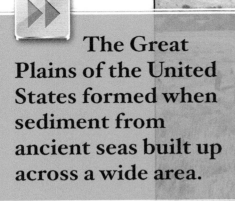

The Great Plains of the United States formed when sediment from ancient seas built up across a wide area.

Plains form in different ways. Some develop by coasts when rivers slow down as they meet the sea. The tiny pieces of rock the rivers erode stop moving, drop, and pile up in flat plains surrounding the river. Other plains form when seas or lakes fill with eroded sediment that has washed or blown off surrounding land.

Plateaus

Plateaus are high, flat areas of land that range in size from a few yards to hundreds of miles wide. Some plateaus formed when land was pushed upward. The Tibetan Plateau rose when two enormous slabs of rock underground pushed into each other, forming a flat, raised area.

Compare and Contrast
How are plateaus and plains similar? How is the height of the surrounding land different?

▶▶ The Tibetan Plateau is the largest in the world. It is about four times the size of Texas!

The mesas of Monument Valley are the remains of an ancient plateau.

Other plateaus formed when layer upon layer of lava built up. Many plateaus formed when existing mountains and plains were eroded at their tops or edges. Flat-topped hills with steep sides are called mesas. Smaller plateaus are called buttes. There are many mesas and buttes in the southwestern United States, such as in Monument Valley.

Mountains

Mountains are the steepest, highest landforms on Earth. Many have a pointed or jagged peak. Some are in rows called **mountain ranges**. The Himalaya mountain range includes Mount Everest. Its peak is more than 5 miles (8 km) high. It formed when underground slabs of rock collided, making surface rock crumple into folds.

Most of the Himalayas lie within China, India, Nepal, and Bhutan.

Mountain ranges are lines of mountains that are connected by high ground.

If this volcanic mountain in the Andes erupts again and more lava flows out of it, the mountain may grow taller.

Some mountains form above cracks in Earth's surface. The rock on one side of the crack can rise above the other. Other mountains form when volcanoes erupt many times and layers of hard lava build up. The highest mountain on Earth is a volcanic mountain called Mauna Kea in Hawaii. It is more than 6 miles (9.6 km) from its base at the ocean floor to its peak.

Hills

Hills are another raised landform. They are not as high as mountains, though there is no official difference between the two.

Some hills formed when rocks collapsed or when blocks of land rose. Others grew when magma, or other liquid masses, pushed up from under the ground, causing the land to bulge.

Compare and Contrast
How are hills and mountains similar? How are they different?

▶▶ The Black Hills of South Dakota formed when a huge underground cave, filled with salty water, pushed up on the rock above it.

Erosion on hills changed the shape and size of these landforms.

Hills also form by erosion. Sometimes ancient mountain rock is broken down over thousands of years by the action of ice forming on rock and then cracking it open. Rain and wind carry pieces of rock away and the mountain shrinks. Sometimes new hills can form when the eroded pieces pile up, too.

Glaciers

Although glaciers themselves are not landforms, they create landforms. Glaciers are large areas of thick ice that remain frozen from one year to the next. Glaciers also slowly flow over the land. They exist all over the world, from Antarctica to mountainous points along the equator.

Glaciers form when layers of snow and ice collect in gaps between high mountains and freeze into a solid lump of ice.

> **Think About It**
>
> The speed at which glaciers move depends on how much water they slide along on. How does a warmer climate affect glaciers?

Glaciers move slowly downhill on water melted from their lowest layers of ice.

Glaciers wear down mountains. Their edges freeze onto and then pull off chunks of rock as they pass. They push along rock and soil in front of them. The front of a glacier melts when it reaches warmer places, leaving piles of rock and soil there!

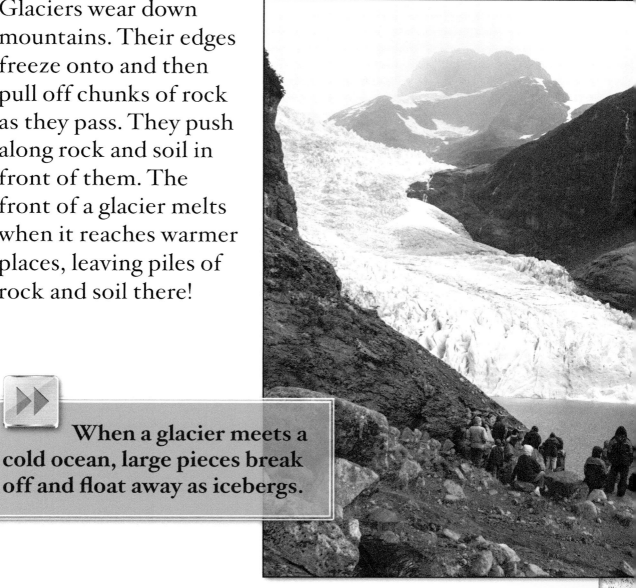

⏩ When a glacier meets a cold ocean, large pieces break off and float away as icebergs.

Valleys

Valleys are long stretches of lower land, usually between hills or mountains. Most valleys form where water wears away rock or soil.

Rainwater runs off slopes into grooves, forming streams. Streams wash away soil, making a narrow, shallow valley. Over time, the streams turn into rivers, with more water to erode sediment from the valley sides and base. The valley then gets deeper and wider.

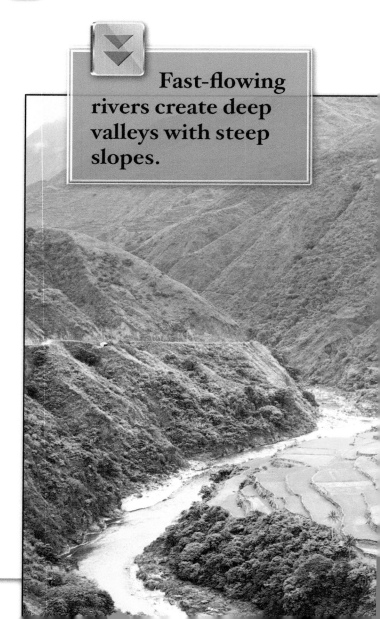

Fast-flowing rivers create deep valleys with steep slopes.

COMPARE AND CONTRAST

How are the shapes of valleys made by rivers and those made by glaciers similar? How are they different?

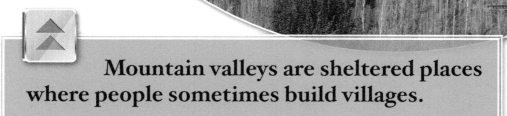

Mountain valleys are sheltered places where people sometimes build villages.

In cold places, glaciers can form in river valleys. They erode these valleys into wider glacial valleys. Other valleys, such as the Rift Valley in Africa, formed when a section of Earth's rock dropped lower than the surrounding land.

Canyons

Canyons are some of the most spectacular valleys on Earth. They are especially deep, narrow valleys often with steep, high sides. One of the biggest of them all is the Grand Canyon. This remarkable landform has taken six million years to form. It is the result of the Colorado River eroding the Colorado Plateau.

>> The Grand Canyon is 277 miles (446 km) long. It is more than a mile (1.6 km) deep at its deepest point.

Canyons with very steep sides are called gorges. They form where rivers carve through weak areas of rock.

Gorges are narrow valleys with steep walls. They often have a stream that runs through them.

Sand Dunes

Sand is made up of tiny pieces of rock eroded from large rocks. Sand is light, so wind can blow it around. In places with lots of sand, such as big beaches or deserts, wind piles sand up into hills called sand dunes. The biggest sand dunes are hundreds of yards high.

Dunes are landforms made up of shifting sand.

Think About It
Some coastal dunes form only where there are plants growing. How do you think these help the sand to pile up?

Sand dunes are constantly changing shape as wind moves sand over and around other dunes.

Sand dunes grow and change over time. The side facing the wind has a gradual slope. It builds up as wind moves the sand grains on top of other grains. Once the dune becomes too steep, one side may fall, leaving a steeper slope. Dunes can move many feet each year, depending on how strong the winds are.

Caves

Caves are underground landforms. They are holes that usually form in limestone. First, rainwater gets underground by soaking into rocks and through cracks. Then it wears away limestone, making gaps that widen over time to form caves. Caves range in size from tiny passages to huge systems of connected "rooms" and tunnels. The world's longest cave system is in Kentucky. It is more than 350 miles (560 km) long.

Water can collect in underground caves, forming pools or even rivers.

Inside limestone caves, there are many special features formed by minerals in water. Minerals in water that drips into the cave can build up into shapes called stalactites, which hang from the roof. Other forms, called stalagmites, grow up where the water splashes onto the floor!

Stalactites are rocky columns that grow down from a cave roof. **Stalagmites** are rocky pillars that grow up from a cave floor.

Most stalactites grow very slowly. It can take 200 years for some to grow an inch (2.5 cm).

COASTLINES

Coastlines are the places where land meets sea. They are shaped by waves. Waves hit rocks with great force, break off pieces, and wash them away. They force air into cracks and split open rocks. Waves also break rocks when they toss sand and stones against them. Waves erode coastlines to make features such as cliffs and coves, which look like giant bites in the coastline.

Waves break pieces from the bottom of cliffs. Then rock falls down from above.

THINK ABOUT IT
The shape of coastal landforms depends on how hard the rock is. Why do you think a bay forms in an area of soft rock?

Waves carry sand onto coastlines and deposit it there, creating beaches.

Coastlines are also formed when waves drop the sediment they are carrying at the edge of land. Beaches form in bays where sand and broken shells collect. Spits are long ridges of sand or rock pieces that stick out from the coastline into the sea.

Islands

Islands are areas of land completely surrounded by water. They can be very small or as large as Greenland and Australia. Islands form in different ways. Some islands were once attached to a larger area of land but were separated by land erosion. Other islands are volcanoes that erupted underwater. Gradually, the lava builds up until the top of the volcano reaches above the water to form an island.

A group of islands close together is an archipelago.

⏩ **St. Michael's Mount in England is an island only when the tide is in.**

A tide is the rising and falling of the ocean.

Some islands are only islands for part of the day! When the tide is out, you can walk across land to these islands. When the tide comes in again, water covers the land between the two, and the island is completely surrounded by water.

Changing Landforms

Landforms are always changing. Some change very slowly, over thousands of years, as they erode. Others change more quickly. People change landforms when they use machines to dig holes in Earth's surface to get coal or other useful substances. The rocks and soil they dig out are often piled up to create new hills, too.

Mining creates new landforms, such as huge holes in Earth's surface.

Think About It

Why do you think people can change landforms more quickly today than they could in the past?

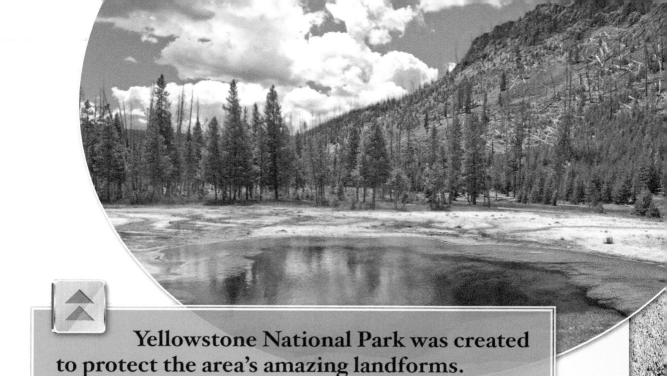

Yellowstone National Park was created to protect the area's amazing landforms.

Sometimes people create new land. The southern tip of Manhattan, New York City, was extended by landfill to increase the size of the island. In the Netherlands, water was pumped out of marshy land so people could live and farm on it. People also protect landforms to stop them from changing. National parks are places where people are not allowed to alter the landforms.

Glossary

Antarctica The very cold region around Earth's South Pole.
archipelago A group of islands that is close together.
beaches Landforms that are created in bays where sand and broken shell pieces collect.
buttes Isolated hills with steep, often vertical sides and flat tops. Buttes are smaller than mesas.
cliffs A high, steep surface of rock, earth, or ice.
coal A fuel burned to create electricity.
collided Pushed into each other.
continent A large, solid area of land.
coves Small, sheltered areas that are set back from the shoreline.
desert A dry area of land that gets very little rain, and as a result, few plants grow there.
eruption When a volcano explodes and hot, melted rock, called lava, spurts out of it.
grassland An area of nearly treeless land covered mainly in grass.
ice Frozen water.
landfill Waste that is buried between layers of earth.
lava Melted rock from inside Earth that comes to the surface when a volcano erupts.
limestone A type of soft rock that erodes easily.
magma Melted rock that is found below Earth's surface.
mesas Flat-topped hills or small plateaus with steep sides.
peak The top of a mountain or hill.
sediment Tiny pieces of rock, mud, dust, or sand.
spits Small points of land that run out into a body of water.
volcanoes Vents in Earth's surface from which melted rock, called lava, spurts out.

For More Information

Books

Ganeri, Anita. *Canyon Hunters* (Landform Adventurers). North Mankato, MN: Raintree, 2012.

Granger, Ronald. *Exploring Earth's Surface* (Exploring Earth and Space). New York, NY: PowerKids Press, 2012.

Kalman, Bobbie. *Introducing Landforms* (Looking at Earth). New York, NY: Crabtree Publishing Company, 2008.

Rau, Dana Meachen. *U.S. Landforms* (True Books: U.S. Regions). Chicago, IL: Children's Press, 2012.

Rosenberg, Pam. *Cave Crawlers* (Landform Adventurers). North Mankato, MN: Raintree, 2012.

Websites

Due to the changing nature of Internet links, Rosen Publishing has developed an online list of Websites related to the subject of this book. This site is updated regularly. Please use this link to access the list:

http://www.rosenlinks.com/lfo/landf

Index

Africa 6, 17
Australia 5, 26

beaches 20, 25
buttes 9

canyons 18–19
    Grand Canyon 18
caves 4, 12, 22–23
cliffs 24
coal 28
coastlines 24–25
coasts 7, 24, 25, 27
coves 24

Earth 4, 6, 10, 11, 17, 18, 28
erosion 5, 7, 9, 13, 15, 16, 17, 18, 20, 24, 26, 28

glaciers 14–15, 17
gorges 19
grassland 6
Greenland 26

Hawaii 11
hills 4, 9, 12–13, 16, 20, 28
    Black Hills 12

ice 5, 13, 14
icebergs 15
islands 4, 26–27, 29
    Manhattan 29
    St. Michael's Mount 27

lava 5, 9, 11, 26
limestone 22, 23

magma 12
mesas 9
    Monument Valley 9
mountains 4, 5, 9, 10–11, 12, 13, 14, 15, 16
    Andes 11
    Himalaya 10
    Mauna Kea 11
    Mount Everest 10

national parks 4, 29
    Bryce Canyon 4
    Yellowstone 29

plains 6–7, 8, 9
    Great Plains 7
    plains animals 6
plateaus 8–9, 18
    Colorado Plateau 18
    Tibetan Plateau 8

rivers 7, 16, 17, 18, 19, 22
    Colorado River 18
rock 4, 5, 7, 8, 10, 11, 12, 13, 15, 16, 17, 19, 20, 22, 23, 24, 25, 28

sand 4, 5, 20, 21, 24, 25
sand dunes 20–21
sea 7, 24, 25
sediment 7, 16, 25
stalactites 23
stalagmites 23
streams 16, 19

tide 27

valleys 16–17, 18, 19
    Rift Valley 17

water 4, 5, 12, 14, 16, 22, 23, 26, 27, 29
waterfall 5
waves 24, 25
wind 5, 13, 20, 21